215
TWA

Twain, Bubba.
The story of our life

The STORY of OUR LIFE,
based on a TRUE LIFE.

The STORY of OUR LIFE, based on a TRUE LIFE.

Bubba "His X. Mark" Twain

XULON PRESS

Xulon Press
2301 Lucien Way #415
Maitland, FL 32751
407.339.4217
www.xulonpress.com

Printed in the United States of America.

ISBN-13: 978-1-54566-536-7

Table of Content:

Foreword by Norma Ramos Quintero:

James N. Akins, Jr. has traveled and worked in more than 40 countries all over the world with the United States government. A man well-traveled certainly has a lot to say, and in his story, *The Message in a Bottle: Norma, I Now Know*, James captures his true-life close to death experience and vividly lets the reader in on some flashbacks and stream of conscience moments that will keep you reading till the last word.

I dedicate this book to God, Christ, the Holy Spirit Mary, Mother of Jesus and my angels that saved my life with their words of truth and support: Dr. Ann Loudon Severance, Norma Ramos Quintero, Dr. Stephen E. Winston, MD and Jeffery J. Pappas, MSW, LICSW, PIP.

Every Christmas I will donate 30% of my book royalties equally to each of the following charities:

- Breast Cancer Research Foundation (New York, NY)

- American Foundation for Suicide Prevention (New York, NY)

- Wounded Warriors Family Support (Omaha, NE)

- St. Jude Children's Research Hospital (Memphis, TN)

- Shriners Hospitals for Children (Tampa, FL)

- Doctors Without Borders (New York, NY)

THE STORY OF OUR LIFE, Based on a True Story –

THE MESSAGE IN A BOTTLE - NORMA, I NOW KNOW

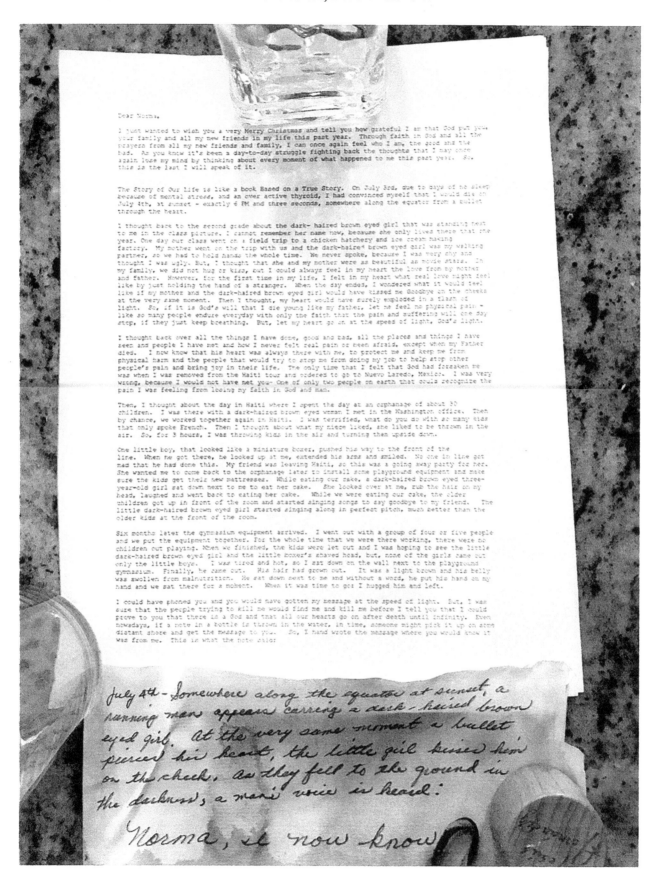

Dear Norma,

I just wanted to wish you a very Merry Christmas and tell you how grateful I am that God put you, your family and all my new friends in my life this past year. Through faith in God and all the prayers from all my new friends and family, I can once again feel who I am, the good and the bad. As you know it's been a day-to-day struggle fighting back the thoughts that I may once again lose my mind by thinking about every moment of what happened to me this past year. So, this is the last I will speak of it.

The Story of Our Life is like a book Based on a True Story. On July 3rd, due to days of no sleep because of mental stress, and an over active thyroid, I had convinced myself that I would die on July 4th, at sunset - exactly 6 PM and three seconds, somewhere along the equator from a bullet through the heart.

I thought back to the second grade about the dark- haired brown eyed girl that was standing next to me in the class picture. I cannot remember her name now, because she only lived there that one year. One day our class went on a field trip to a chicken hatchery and ice cream making factory. My mother went on the trip with us and the dark-haired brown eyed girl was my walking partner, so we had to hold hands the whole time. We never spoke, because I was very shy and thought I was ugly. But, I thought that she and my mother were as beautiful as movie stars. In my family, we did not hug or kiss, but I could always feel in my heart the love from my mother and father. However, for the first time in my life, I felt in my heart what real love might feel like by just holding the hand of a stranger. When the day ended, I wondered what it would feel like if my mother and the dark-haired brown eyed girl would have kissed me Goodbye on the cheeks at the very same moment. Then I thought, my heart would have surely exploded in a flash of light. So, if it is God's will that I die young like my father, let me feel no physical pain - like so many people endure everyday with only the faith that the pain and suffering will one day stop, if they just keep breathing. But, let my heart go on at the speed of light, God's light.

I thought back over all the things I have done, good and bad, all the places and things I have seen and people I have met and how I never felt real pain or been afraid, except when my Father died. I now know that his heart was always there with me, to protect me and keep me from physical harm and the people that would try to stop me from doing my job to help stop other people's pain and bring joy in their life. The only time that I felt that God had forsaken me was when I was removed from the Haiti tour and ordered to go to Nuevo Laredo, Mexico. I was very wrong, because I would not have met you. One of only two people on earth that could recognize the pain I was feeling from losing my faith in God and man.

Then, I thought about the day in Haiti where I spent the day at an orphanage of about 30 children. I was there with a dark-haired brown eyed woman I met in the Washington office. Then by chance, we worked together again in Haiti. I was terrified, what do you do with so many kids that only spoke French. Then I thought about what my niece liked, she liked to be thrown in the air. So, for 3 hours, I was throwing kids in the air and turning them upside down.

One little boy, that looked like a miniature boxer, pushed his way to the front of the line. When he got there, he looked up at me, extended his arms and smiled. No one in line got mad that he had done this. My friend was leaving Haiti, so this was a going away party for her. She wanted me to come back to the orphanage later to install some playground equipment and make sure the kids get their new mattresses. While eating our cake, a dark-haired brown eyed three-year-old girl sat down next to me to eat her cake. She looked over at me, rub the hair on my head, laughed and went back to eating her cake. While we were eating our cake, the older children got up in front of the room and started singing songs to say goodbye to my friend. The little dark-haired brown eyed girl started singing along in perfect pitch, much better than the older kids at the front of the room.

Six months later the gymnasium equipment arrived. I went out with a group of four or five people and we put the equipment together. For the whole time that we were there working, there were no children out playing. When we finished, the kids were let out and I was hoping to see the little dark-haired brown eyed girl and the little boxer's shaved head, but, none of the girls came out only the little boys. I was tired and hot, so I sat down on the wall next to the playground gymnasium. Finally, he came out. His hair had grown out. It was a light brown and his belly was swollen from malnutrition. He sat down next to me and without a word, he put his hand on my hand and we sat there for a moment. When it was time to go; I hugged him and left.

I could have phoned you and you would have gotten my message at the speed of light. But, I was sure that the people trying to kill me would find me and kill me before I tell you that I could prove to you that there is a God and that all our hearts go on after death until infinity. Even nowadays, if a note in a bottle is thrown in the water, in time, someone might pick it up on some distant shore and get the message to you. So, I hand wrote the message where you would know it was from me. This is what the note said:

July 4th - Somewhere along the equator at sunset, a running man appears carrying a dark-haired brown eyed girl. At the very same moment a bullet pierced his heart, the little girl kissed him on the check. As they fell to the ground in the darkness, a man's voice is heard:

Norma, I now know

Chapter 1

THE MESSAGE IN A BOTTLE:

NORMA, I NOW KNOW

<u>RADIO ANNOUNCER</u>–Happy New Years. I have a special request to read a note found on a Gulf of Mexico beach in a tequila bottle.

Dear Norma,

I just wanted to wish you a very Merry Christmas and tell you how grateful I am that God put you, your family and all my new friends in my life this past year. Through faith in God and all the prayers from all my new friends and family, I can once again feel who I am, the good and the bad. As you know it's been a day-to-day struggle fighting back the thoughts that I may once again lose my mind by thinking about every moment of what happened to me this past year. So, this is the last I will speak of it.

The Story of Our Life is like a book Based on a True Story. On July 3rd, due to days of no sleep because of mental stress, and an over active thyroid, I had convinced myself that I would die on July 4th, at sunset–exactly 6 PM three seconds, somewhere along the equator from a bullet through the heart.

I thought back to the second grade about the dark-haired brown eyed girl that was standing next to me in the class picture. I cannot remember her name now, because she only lived there that one year. One day our class went on a field trip to a chicken hatchery and ice cream making factory. My mother went on the trip with us and the dark-haired brown eyed girl was my walking partner, so we had to hold hands the whole time. We never spoke, because I was very shy and thought I was ugly. But, I thought that she and my mother were as beautiful as movie stars. In my family, we did not hug or kiss, but I could always feel in my heart the love from my mother and father. However, for the first time in my life, I felt in my heart what real love might feel like by just holding the hand of a stranger. When the day ended, I wondered what it would feel like if my mother and the dark-haired brown eyed girl would have kissed me Goodbye on the cheeks at the very same moment. Then I thought, my heart would have surely exploded in a flash of light. So, if it is God's will that I die young like my father, let me feel no physical pain–like so many people endure everyday with only the faith that the pain and suffering will one day stop, if they just keep breathing. But, let my heart go on at the speed of light, God's light. I thought back over all the things I have done, good and bad, all the places and things I have seen and people I have met and how I never felt real pain or been afraid, except when my Father died. I now know that his heart was always there with me, to protect me and keep me from physical harm and the people that would try to stop me from doing my job to help stop other people's pain and bring joy in their life. The only time that I felt that God had forsaken me was when I was removed from the Haiti tour and ordered to go to Nuevo Laredo, Mexico. I was very wrong, because I would not have met you– One of only two people on earth that recognize the pain I was feeling from losing my faith in God and man.

Then, I thought about the day in Haiti where I spent the day at an orphanage of about 30 children. I was there with a dark-haired brown eyed woman I met in the Washington office. Then by chance, we worked together again in Haiti. I was terrified, what do you do with so many kids that only spoke French. Then I thought about what my niece liked, she liked to be thrown in the air. So, for 3 hours, I was throwing kids in the air and turning them upside down.

One little boy, that looked like a miniature boxer, pushed his way to the front of the line. When he got there, he looked up at me, extended his arms and smiled. No one in line got mad that he had done this. My friend was leaving Haiti, so this was a going away party for her. She wanted me to come back to the orphanage later to install some playground equipment and make sure the kids get their new mattresses. While eating our cake, a dark-haired brown eyed three-year-old girl sat down next to me to eat her cake. She looked over at me, rub the hair on my head, laughed and went back to eating her cake. While we were eating our cake, the older children got up in front of the room and started singing songs to say goodbye to my friend. The little dark-haired brown eyed girl started singing along in perfect pitch, much better than the older kids at the front of the room.

Six months later the gymnasium equipment arrived. I went out with a group of four or five people and we put the equipment together. For the whole time that we were there working, there were no children out playing. When we finished, the kids were let out and I was hoping to see the little dark-haired brown eyed girl and the little boxer's shaved head, but, none of the girls came out only the little boys. I was tired and hot, so I sat down on the wall next to the gymnasium. Finally, he came out. His hair had grown out. It was a light brown and his belly was swollen from malnutrition. He sat down next to me and without a word, he put his hand on my hand and we sat there for a moment. When it was time to go: I hugged him and left.

I could have phoned you and you would have gotten my message at the speed of light. But, I was sure that the people trying to kill me would find me and kill me before I tell you that I could prove to you that there is a God and that all our hearts go on after death until infinity. Even nowadays, if a note in a bottle is thrown in the water, in time, someone might pick it up on some distant shore and get the message to you. So, I hand wrote the message where you would know it was from me. This is what the note said:

July 4th, somewhere along the equator at sunset, a running man appears carrying a dark-haired brown eyed girl. At the very same moment a bullet pierces his heart, the little girl kisses him on the cheek. As they fell to the ground in darkness, a man's voice is heard: Norma, I now Know

RADIO ANNOUNCER–This note reminds me that my brother once accidently drank a full bottle of tequila toasting colleagues, alive and dead, and ended up imagining he could exceed the speed of light, at least in the first quarter mile.

Well, this song goes out to all our Brothers and Sisters that have put themselves in harm's way to make our life better, and will not make it home this New Year. On a personal note, I want to thank God and his sweet Angels for bringing my Bubba home safe this past year.

Our next song is from Norma, our favorite beautiful Dark-haired Brown Eyed girl singing her latest hit: *My Heart will go on.*

Song–Every night in my dreams, I see you, I feel you, that is how I know you go on.

Chapter 2

NORMA, I NOW KNOW what happens to Mass if it Exceeds the Speed of Light.

In the YouTube video, you sent me; There Is A God–Lee Ann Womack, there is a picture of Albert Einstein. When I viewed the video, I had not slept for a week and I was thinking about nothing but, how to explain to you; "What happens to mass if it exceeds the speed of light?" I thought, how could God let Albert tell the world the scientific formula for the Atom Bomb and how could he ever smile again after he saw the how many humans died and the subsequent "Cold War" that followed [Albert died in 1955 the year after I was born.] The whole world believed Albert Einstein's formula "E=mc² " was the key to how the Atom Bomb works. Before they set off the first Atom Bomb at White Sands, New Mexico [I worked a few miles from there in Alamogordo, NM.] the scientist thought the Atom Bomb would cause a chain reaction in the atmosphere and destroy earth. Well, the world did not blow up, but more than 115 thousand humans were killed or deformed from the two bombs that were dropped on Japan [Fortunately, millions of humans were saved by convincing the Emperor of Japan to stop the War.]

So, I broke Einstein's formula down by defining each of the letters and making a sentence out of them to see if the formula made any since. E [ALL THE ENERGY IN THE UNIVERSE] = [EQUALS] m [ALL THE MASS IN THE UNIVERSE] (In physics, mass is a property of a physical body, is often seen as spherical because it is such a fundamental quantity that it is hard to define in terms of something else.) times c^2 [c = CONSTANT SPEED OF LIGHT times c = CONSTANT SPEED OF LIGHT.]

The first of two atom bombs that exploded over Japan was called "Little Boy," its design used the gun method to explosively push 38.5 (85 pounds) kilograms hollow cylinder of sub-critical mass of uranium-235 1.100 meters (42 inches), by means of four cylindrical silk bags of cordite (Burn rate 1,000 meter/second), into a "target" 25.6 (56 pounds) kilogram solid target cylinder together into a super-critical mass, initiating a nuclear chain reaction.

To this date, physicist express "E = mc² " as the dynamic equation for calculating FORCE to measure the ENERGY of the atom bombs. So, now the sentence reads:E[FORCE = JOULES (38,500,000-kilograms times meter² per second²] =[EQUALS] m[38.5 kilograms mass] times c²= a (acceleration)[1,000 meters per second times 1,000 meters per second or 1,000,000-meter² per second².]

38,500,000 Joules is a far cry less than 38.5 kilograms mass times c²(Constant Speed of Light² = 89,875,517,873,681,800-meter²/second²) or 3,460,207,438,000,000,000Joules.

As anyone can see E = mc² is not the same Dynamic equation as F = ma, because no explosives on earth can propel a mass into another mass or implode one mass into another mass at 89,875,517,873,681,800-meter²/second².

Mass has two states of existence; Stop or Static and Moving or Dynamic.

If one simply changes the Dynamic Force and E=mc² equations to Static equations by removing the movement of acceleration and Constant Speed of Light² the Force equation would read; mass (kilograms) = mass (kilograms) and Einstein's equation would read; ENERGY = MASS (E=M), therefore, it is a false assumption that the two Dynamic equations are equal.

Did Einstein lie to the scientific community? I do not think so, because he used a small letter "m" in his formula to represent a quantity of mass less than "ALL THE MASS IN THE UNIVERSE". No doubt Einstein was a genius, but any high school math student knows the difference between a Static equation and a Dynamic equation and how to check one's work. I think Albert found the equation in some secret documents the Germans stole from one of the countries they invaded and he did not know what it meant. So, "Finders keepers. Losers weepers." Norma, I now know what E = Mc² (Big letter "M") means.

"If you knew the magnificence of the three, six and nine, you would have a key to the universe." – Nikola Tesla

Even Albert Einstein did not know, although his formula E=Mc² clearly defined the magnificence of 3, 6, and 9. His formula could also be written: (E=M) [rotating sphere of magnetic spheres] = ½ (E=M) (c) · ½ (E=M) (c). Also, I have mathematical proof that Pi = 3 not 3.14.

"Key to the universe: The universe, just like the earth, is not flat." – James N. Akins, Jr.

The information in Chapter 3, may lead to more efficient power generation and electric motors, faster computers, more efficient computer operating systems and memory storage, Big Bang Law, String Law, and Law of General Relativity, cure for cancer, define Black Holes, fine more Goldilocks planets and how to communicate with other planets.

Norma, I now know it is God's law that no mass may exceed the speed of light and Albert Einstein could smile and be happy, because of his faith in God's love for him and the lives he did help save during such a horrific time in human history.

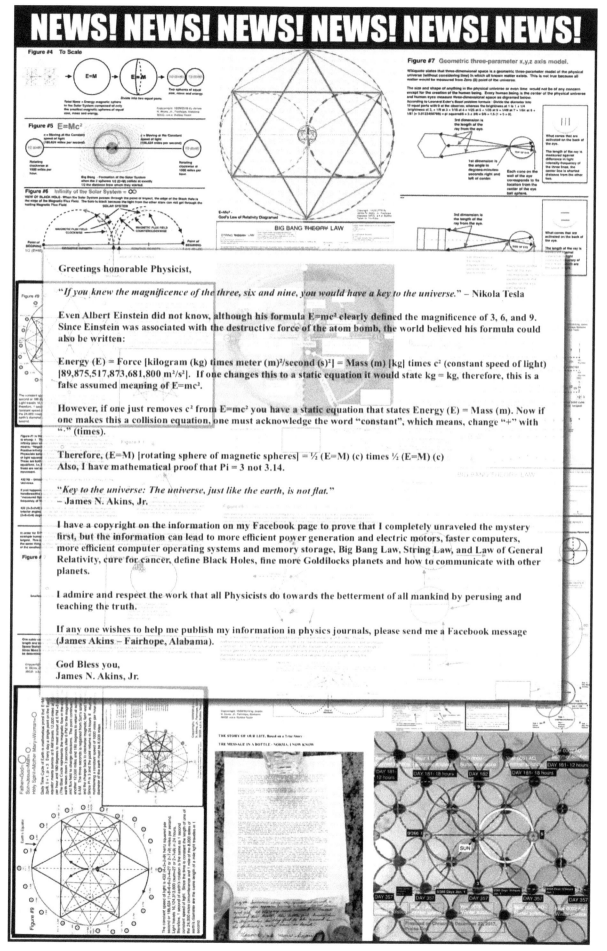

Greetings honorable Physicist,

"If you knew the magnificence of the three, six and nine, you would have a key to the universe." – Nikola Tesla

Even Albert Einstein did not know, although his formula E=mc² clearly defined the magnificence of 3, 6, and 9. Since Einstein was associated with the destructive force of the atom bomb, the world believed his formula could also be written:

Energy (E) = Force [kilogram (kg) times meter (m)²/second (s)²] = Mass (m) [kg] times c² (constant speed of light) [89,875,517,873,681,800 m²/s²]. If one changes this to a static equation it would state kg = kg, therefore, this is a false assumed meaning of E=mc².

However, if one just removes c² from E=mc² you have a static equation that states Energy (E) = Mass (m). Now if one makes this a collision equation, one must acknowledge the word "constant", which means, change "+" with "." (times).

Therefore, (E=M) [rotating sphere of magnetic spheres] = ½ (E=M) (c) times ½ (E=M) (c)
Also, I have mathematical proof that Pi = 3 not 3.14.

"Key to the universe: The universe, just like the earth, is not flat."
– James N. Akins, Jr.

I have a copyright on the information on my Facebook page to prove that I completely unraveled the mystery first, but the information can lead to more efficient power generation and electric motors, faster computers, more efficient computer operating systems and memory storage, Big Bang Law, String Law, and Law of General Relativity, cure for cancer, define Black Holes, fine more Goldilocks planets and how to communicate with other planets.

I admire and respect the work that all Physicists do towards the betterment of all mankind by perusing and teaching the truth.

If any one wishes to help me publish my information in physics journals, please send me a Facebook message (James Akins – Fairhope, Alabama).

God Bless you,
James N. Akins, Jr.

7

Chapter 3

I now know..., written by James N. Akins, Jr.

Figure #1 is the present understanding of the meaning of E=mc², it is wrong: 1. The symbol for infinity ∞ and the definition of infinity (also known as Forever) is wrong. The symbol for Infinity means; "Negative Infinity = ∞ = Positive Infinity", whereas, Positive Infinity minus Negative Infinity = Pi (Not Forever). 2. Physicists believe $E=mc^2$ (Energy=mass times the constant speed of light squared) is equal to F=ma (Force=mass times acceleration). These are both Dynamics equations. By making them Static equations, i.e, E(Energy)=m(Mass) is equal to (F)Force = m (mass) these are not equal equations, because Force is just Mass without movement.

432 Hz – Unlocking The Magnificence of the 3 6 9: The Key To The Universe.

It just happens that the constant speed of light is 432 cubic handbreadths squared, or 186,624 miles per second, not the "measured Speed of Light" and verified by Music Standard tuning frequency of Verdi's 'A' = 432 Hz (also, the frequency of red light.)

432 (4+3+2=9) divided by 12 = 36 (3+6= 9) A circle is 360 degrees in interior angles. Divide 360 by 10 parts is 36 degrees each part. 360 (3+6+0=9) degrees divided by 12 (1+2=3) hours is 30 (3+0=3) degrees.

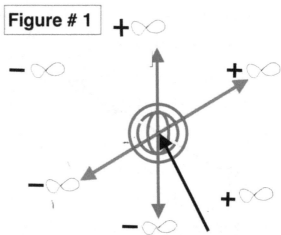

Figure # 1

Present understanding of the Big Bang theory is $E=mc^2$ caused the Big Bang, whereas, there was an explosion that cooled down and formed the Universe into infinity in all directions from 0.

In order for E=M or Energy = Mass they must be the same thing and they must be definite in size not indefinite or unlimited. For example humans believe that in the Universe, it is possible to have something smaller than the smallest and larger than the largest. This is not true in our Solar system. The Universe outside our Solar System is made up of their own $E=Mc^2$'s. Energy is the same thing as Mass because the smallest particle is a magnetic solid sphere that started out as a large solid sphere made up of the smallest magnetic spheres.

Not to scale:

Figure # 2

○

Smallest particle magnetic sphere

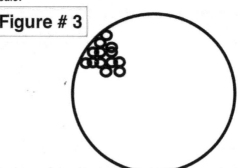

Figure # 3

Total Mass = Energy magnetic sphere in the Solar System composed of only the smallest magnetic spheres of equal size, mass and energy.

One cubic centimeter of water is one gram (Mass). The measurement unit is not significant, only that all sides are equal in length and water's mass is exactly equal to 1. Place 1 gram of water in a vacuum to form a sphere, like in the International Space Station. The actual length of the diameter is 1 measuring unit according to the geometry of a circle and not 1 centimeter. Since Mass is equal to Energy, the energy maybe be measured in 1 gram of water and the diameter of the smallest spheres may be determined.

Figure #4   To Scale

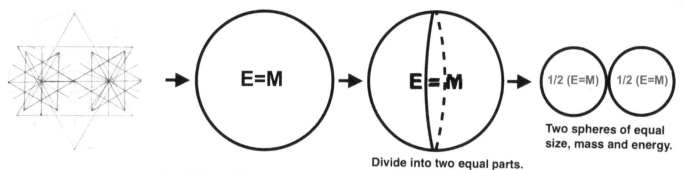

Total Mass = Energy magnetic sphere in the Solar System composed of only the smallest magnetic spheres of equal size, mass and energy.

Divide into two equal parts.

Two spheres of equal size, mass and energy.

Figure #5 $E=Mc^2$

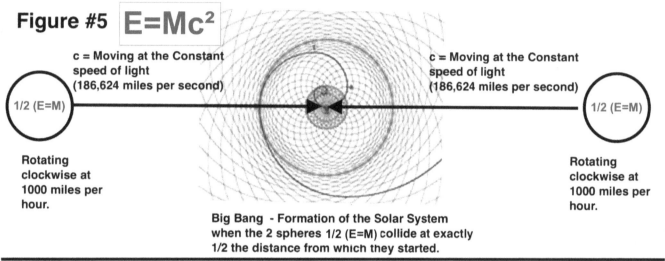

c = Moving at the Constant speed of light (186,624 miles per second)

c = Moving at the Constant speed of light (186,624 miles per second)

1/2 (E=M)

1/2 (E=M)

Rotating clockwise at 1000 miles per hour.

Rotating clockwise at 1000 miles per hour.

Big Bang - Formation of the Solar System when the 2 spheres 1/2 (E=M) collide at exactly 1/2 the distance from which they started.

Figure #6   Infinity of the Solar System = ∞

VIEW OF BLACK HOLE- When the Solar System passes through the point of impact, the edge of the Black Hole is the edge of the Magnetic Flux Field. The hole is black because the light from the other stars can not get through the trailing Magnetic Flux Field.

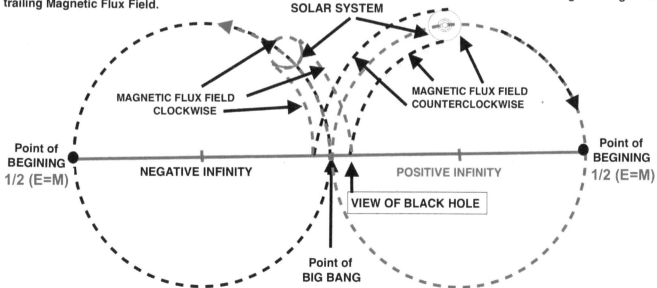

SOLAR SYSTEM

MAGNETIC FLUX FIELD CLOCKWISE

MAGNETIC FLUX FIELD COUNTERCLOCKWISE

Point of BEGINING 1/2 (E=M)

NEGATIVE INFINITY

POSITIVE INFINITY

Point of BEGINING 1/2 (E=M)

VIEW OF BLACK HOLE

Point of BIG BANG

Figure #7   Geometric three-parameter x,y,z axis model.

Wikiquote states that three-dimensional space is a geometric three-parameter model of the physical universe (without considering time) in which all known matter exists. This is not true because all matter would be measured from Zero (0) point of the universe.

The size and shape of anything in the physical universe or even time would not be of any concern except for the creation of the human being. Every human being is the center of the physical universe and human eyes measure three-dimensional space as digramed below.

According to Leonard Euler's Basel problem formula : Divide the diameter into 10 equal parts with 0 at the observer, whereas the brightness at 1 is 1 + 1/4 brightness at 2, + 1/9 at 3 + 1/16 at 4 + 1/25 at 5 + 1/36 at 6 + 1/49 at 7 + 1/64 at 8 + 1/81 (= 0.0123456789) = pi squared/6 = 3 x 3/6 = 9/6 = 1.5 (1 + 5 = 6).

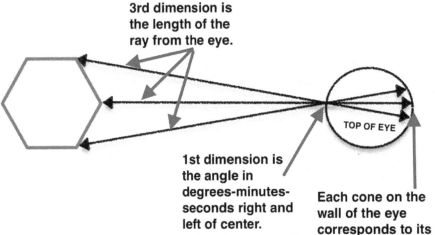

3rd dimension is the length of the ray from the eye.

TOP OF EYE

1st dimension is the angle in degrees-minutes-seconds right and left of center.

Each cone on the wall of the eye corresponds to its location from the center of the eye ball sphere.

What cones that are activated on the back of the eye.

The length of the ray is measured against differance in light intensity-frequency of the three lines, the center line is shorted distance from the other two.

3rd dimension is the length of the ray from the eye.

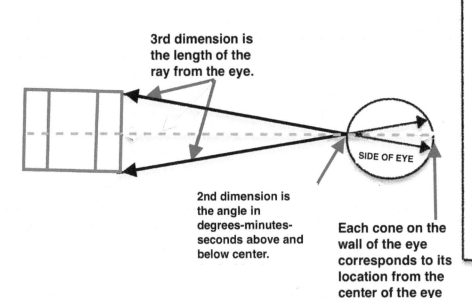

SIDE OF EYE

2nd dimension is the angle in degrees-minutes-seconds above and below center.

Each cone on the wall of the eye corresponds to its location from the center of the eye ball sphere.

What cones that are activated on the back of the eye.

The length of the ray is measured against differance in light intensity-frequency of the 2 lines, which are the same lenght.

Figure #8 $E = Mc^2$:[1 = 0], [2Pi/6], [Pi = 3], [(+) Infinity minus (-) Infinity = Pi]

Referance Figure #6 below, The 'Point of Big Bang" is [0 = 1] because it starts at 0 and goes 360 degrees in the clockwise direction back to the same point that is now 1, then it is 0 again and goes 360 degrees in the counterclockwise direction back to the same point that is 1 again. In both cases it is 1 full revolution equaling a distance of Pi for each circle with a diameter of 1d. Disconnect at 1 = 0 point and move top 1/2 up 1d and bottom 1/2 down 1 d. Divide the 2pi circle into 6 parts as indicated by the blue lines. See diagram below.

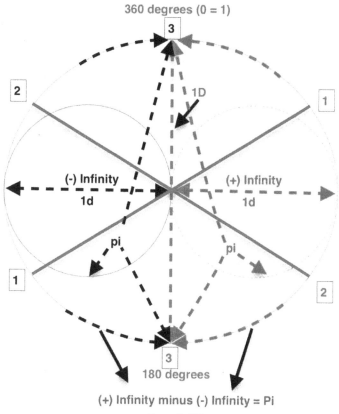

360 degrees (0 = 1)

1D

2

1

(-) Infinity
1d

(+) Infinity
1d

1

2

pi

pi

3

180 degrees

(+) Infinity minus (-) Infinity = Pi

2pi/6

This is the order of the Solar System/Universe and the meaning of $E = Mc^2$. This is mathematical proof that Pi is equal to 3 not 3.14, the diameter of any circle is 1 and is always a ratio of 1 to 3. The actual physical length of the diameter of any circle does not change a circles geomeery because the radius of a circle is always 0.5 the physical length of the diameter and a circle is divided into 360 individual degree of equal length along the circumferance of the circle.

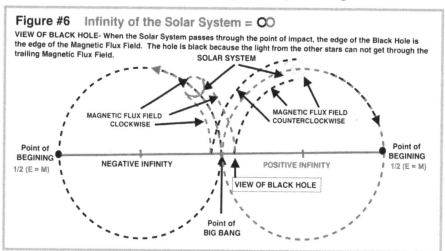

Figure #6 Infinity of the Solar System = ∞

VIEW OF BLACK HOLE- When the Solar System passes through the point of impact, the edge of the Black Hole is the edge of the Magnetic Flux Field. The hole is black because the light from the other stars can not get through the trailing Magnetic Flux Field.

SOLAR SYSTEM

MAGNETIC FLUX FIELD
CLOCKWISE

MAGNETIC FLUX FIELD
COUNTERCLOCKWISE

Point of
BEGINING
1/2 (E = M)

NEGATIVE INFINITY

POSITIVE INFINITY

Point of
BEGINING
1/2 (E = M)

VIEW OF BLACK HOLE

Point of
BIG BANG

Father=God= ◯

Son=Jesus=Man= ◯ ,

Holy Spirit=Mother Mary=Woman=

Daily Time Cycle of Earth is mathmatical proof that E=Mc², 2pi/6, 0 = 1, pi = 3. Every day a single point on the Earth's equator meets sunrise at 6 AM travels 12,000 miles at 1,000 per hour and 180 degrees to meet sunset at 6 PM +3 seconds (The Blue Circle represents the magnetic flow in the earth, the earth slows down 3 seconds after 6 PM for the magnetic flow and flux field to change directions. The point continues another 12,000 miles and 180 degrees to return at sunrise at 6 AM. The three seconds is regained from Sun's solar energy and the change back to clockwise magnetic field and flux field. Since Pi is 3 and the point returns in 24 hours it must be maintaining a constant speed of 1000 miles per hour and the diameter of the earth must be 8,000 miles.

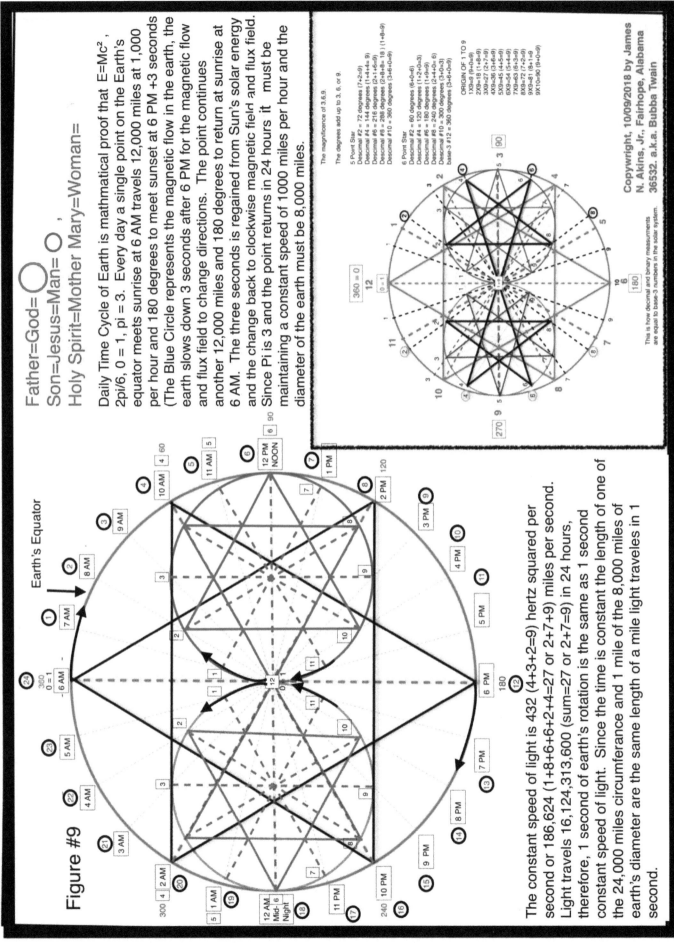

The magnificence of 3.6.9.

The degrees add up to 3, 6, or 9.

5 Point Star
Descimal #2 = 72 degrees (7+2=9)
Descimal #4 = 144 degrees (1+4+4= 9)
Descimal #6 = 216 degrees (2+1+6=9)
Descimal #8 = 288 degrees (2+8+8= 18) (1+8=9)
Descimal #10 = 360 degrees (3+6+0=9)

6 Point Star
Descimal #2 = 60 degrees (6+0=6)
Descimal #4 = 120 degrees (1+2+0=3)
Descimal #6 = 180 degrees (1+9=9)
Descimal #8 = 240 degrees (2+4+0= 6)
Descimal #10 = 300 degrees (3+0=3)
base-3 #12 = 360 degrees (3+6+0=9)

ORIGIN OF 1 TO 9
1X9=9 (9+0=9)
2X9=18 (1+8=9)
3X9=27 (2+7=9)
4X9=36 (3+6=9)
5X9=45 (4+5=9)
6X9=54 (5+4=9)
7X9=63 (6+3=9)
8X9=72 (7+2=9)
9X9=81 (8+1=9)
9X10=90 (9+0=9)

This is how decimal and binary measurments are equal to base-3 numbers in the solar system.

Figure #9

Earth's Equator

The constant speed of light is 432 (4+3+2=9) hertz squared per second or 186,624 (1+8+6+6+2+4=27 or 2+7+9) miles per second. Light travels 16,124,313,600 (sum=27 or 2+7=9) in 24 hours, therefore, 1 second of earth's rotation is the same as 1 second constant speed of light. Since the time is constant the length of one of the 24,000 miles circumferance and 1 mile of the 8,000 miles of earth's diameter are the same length of a mile light traveles in 1 second.

Figure #10 $E=Mc^2$

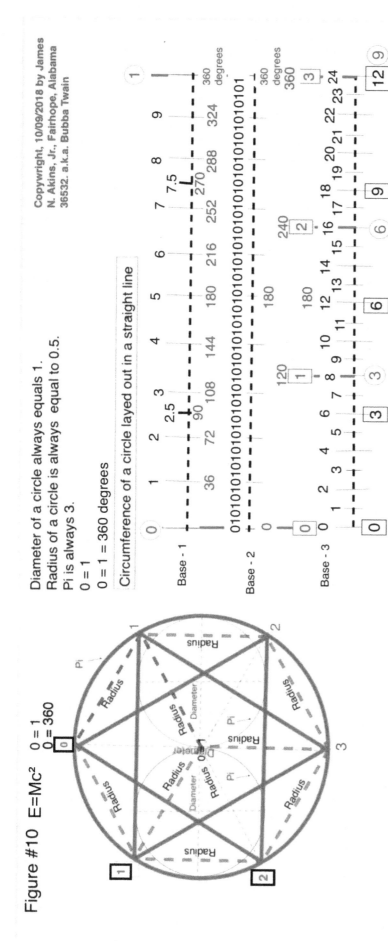

Copywright, 10/09/2018 by James N. Akins, Jr., Fairhope, Alabama 36532. a.k.a. Bubba Twain

Diameter of a circle always equals 1.
Radius of a circle is always equal to 0.5.
Pi is always 3.
0 = 1
0 = 1 = 360 degrees

Circumference of a circle layed out in a straight line

Base - 1

Base - 2

0101010101010101010101010101010101010101010101

Base - 3

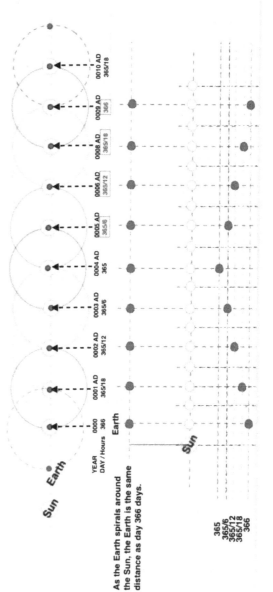

As the Earth spirals around the Sun, the Earth is the same distance as day 366 days.

YEAR DAY / Hours		
0000 366	Earth	

365
365/6
365/12
365/18
366

Note the green line tracing a geometric cube in two dimensions. The third dimention - Reference Figure #7, if this was a physical solid cube your eyes determine the distance of each point on the 3 surfaces by the different frequencies/brightness of green from shortest to longest distance from your eyes.

Combined with the above diagram for $E=Mc^2$, the diagram on right is mathematical proof God created the universe in 6 days (360 days around a circle and 6 days (24 Years) traveling in a spiral.) and rested on the 7th day (4 Years).

Every 4 years on leap year we add 1 day. This is not possible because the Earth would have left the Sun's gravity pull a long time ago.   The earth is able to stay in orbit of the Sun by being pulled 1 day closer to the Sun then resisting its gravity to reach maximum orbit of 366 days every 8 Years.

15

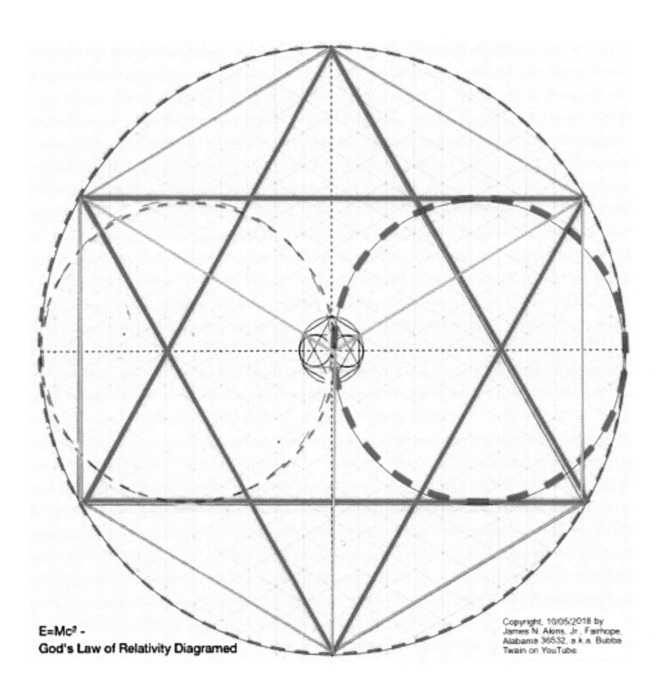

E=Mc² -
God's Law of Relativity Diagramed

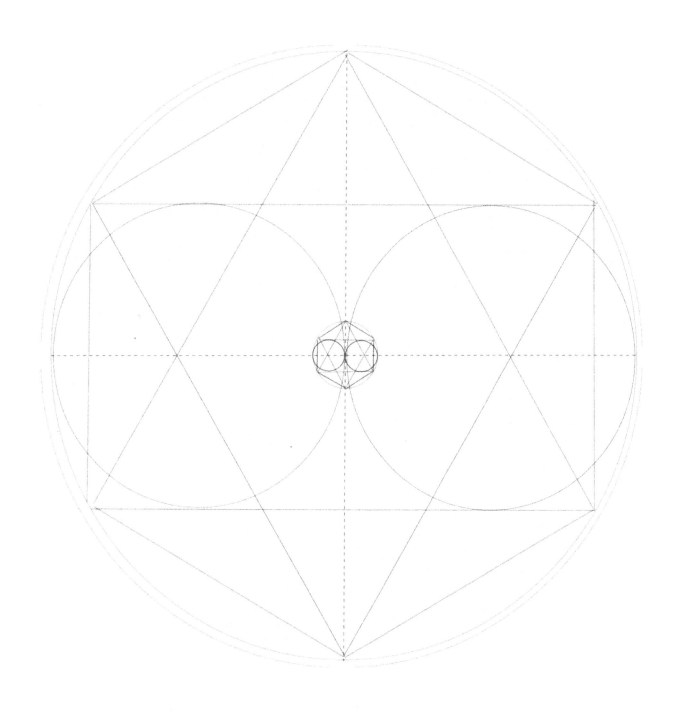

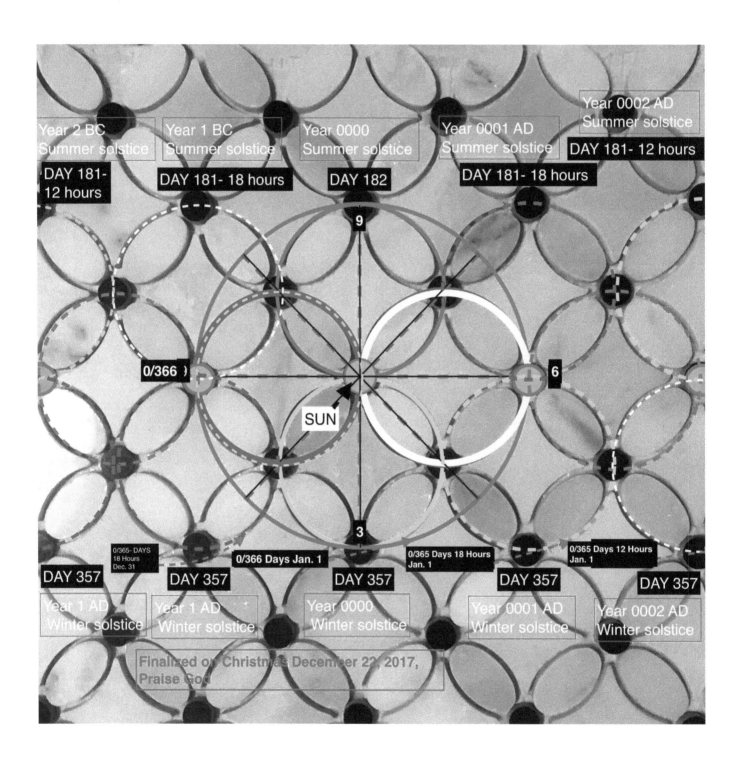

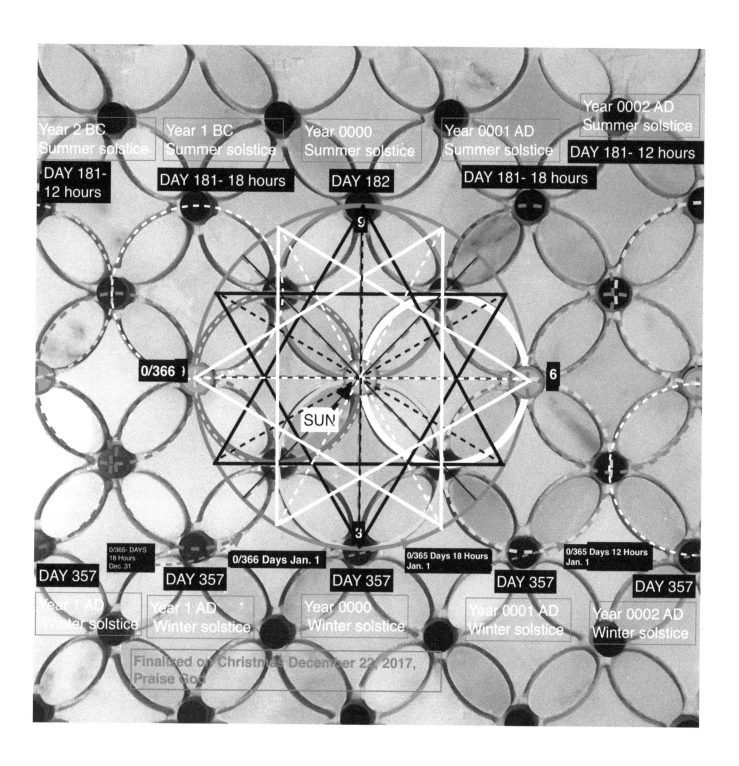

Year 2 BC
Summer solstice

Year 1 BC
Summer solstice

Year 0000
Summer solstice

Year 0001 AD
Summer solstice

Year 0002 AD
Summer solstice

DAY 181-12 hours

DAY 181-
12 hours

DAY 181- 18 hours

DAY 182

DAY 181- 18 hours

0/366

6

SUN

0/365- DAYS
18 Hours
Dec. 31

0/366 Days Jan. 1

0/365 Days 18 Hours
Jan. 1

0/365 Days 12 Hours
Jan. 1

DAY 357

DAY 357

DAY 357

DAY 357

DAY 357

Year 1 AD
Winter solstice

Year 1 AD
Winter solstice

Year 0000
Winter solstice

Year 0001 AD
Winter solstice

Year 0002 AD
Winter solstice

Finalized on Christmas December 25, 2017,
Praise God

19

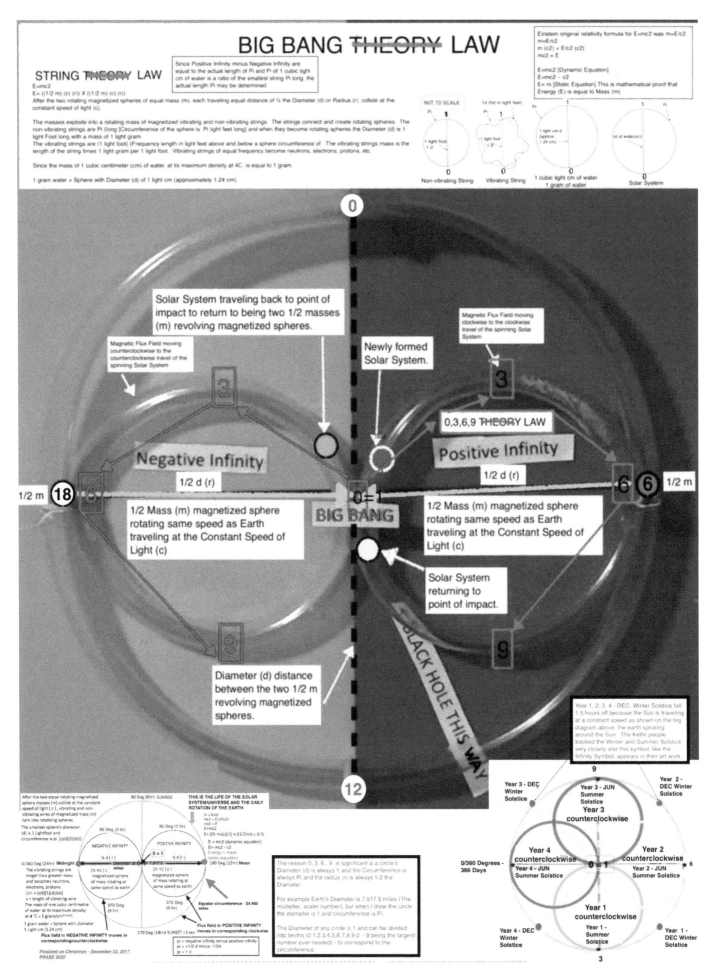

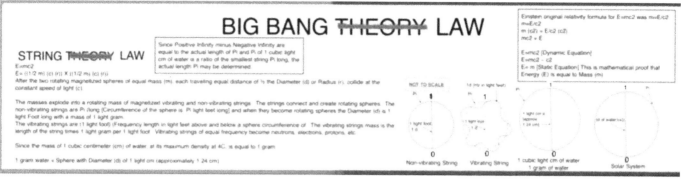

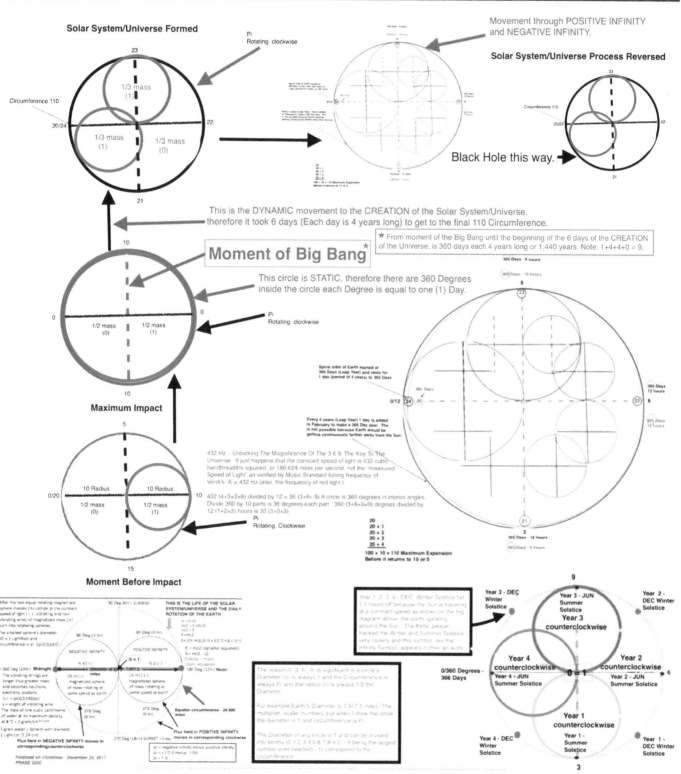

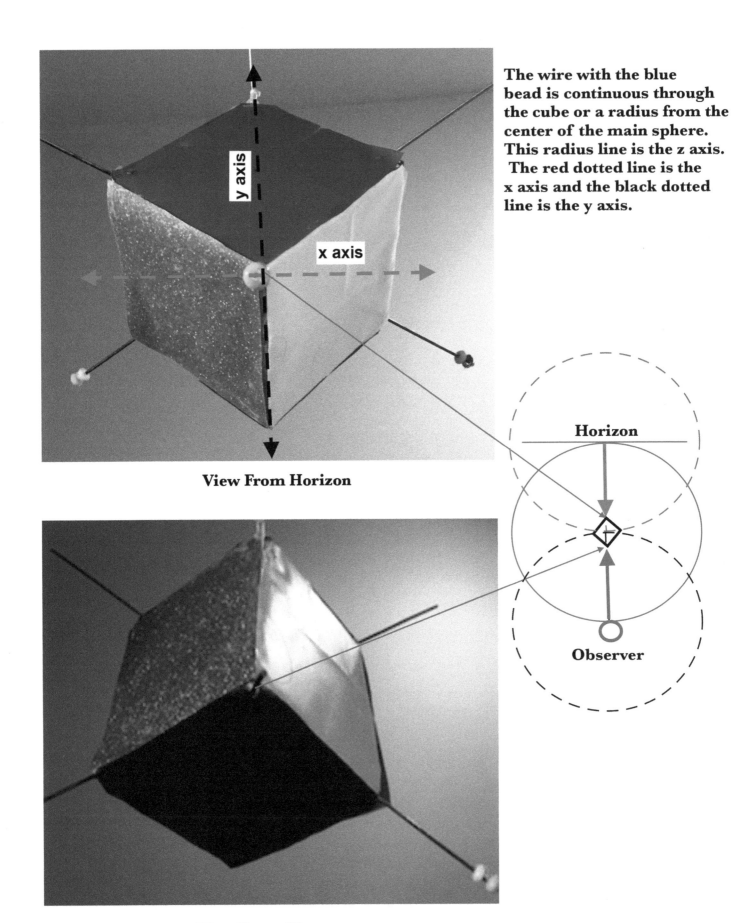

The wire with the blue bead is continuous through the cube or a radius from the center of the main sphere. This radius line is the z axis. The red dotted line is the x axis and the black dotted line is the y axis.

y axis

x axis

View From Horizon

Horizon

Observer

View From Observer

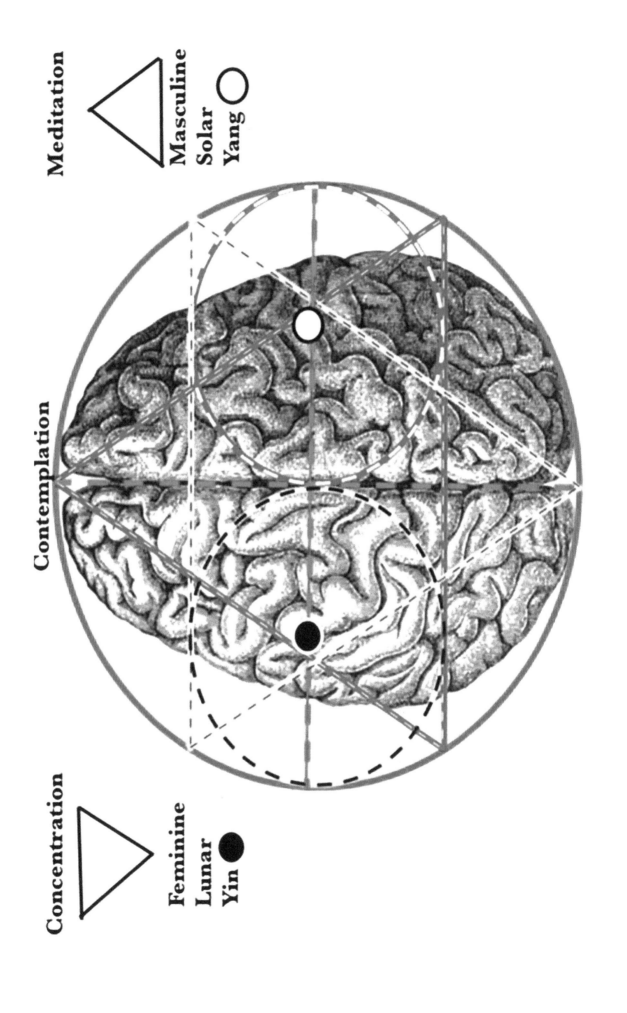

Meditation

Masculine
Solar
Yang ○

Contemplation

Concentration

Feminine
Lunar
Yin ●

The explosion of the Atom Bomb has nothing to do with the equation E = mc2. The equation used is Force [Joules] = Mass [M] x Acceleration [m/s/s].

The explosion of the Atom Bomb is depicted as expanding in the (+) and (-) x, y, z axis, equal to the expansion of the "BIG BANG" that created te Universe and our solar system. This is not correct, everything is based on revolving spheres moving in circles around other spheres. Has anyone seen a cube moving around other objects in a rectangle path.

THIS IS THE LIFE OF THE SOLAR SYSTEM/UNIVERSE AND THE DAILY ROTATION OF THE EARTH

After the two equal rotating magnetized sphere masses (m) collide at the constant speed of light (c) , vibrating and non-vibrating wires of magnatized mass (m) turn into rotateing spheres.

The smallest sphere's diameter (d) is 1 Lightfoot and circumference is pi. [(pi)((1)(d))]

$m = E/c2$
$mc2 = E/c2(c2)$
$mc2 = E$
$E = mc2$
$E = ((½ m)(c)(r)) \times ((1/2m)(c)(r))$

$E = mc2$ [dynamic equation]
$E = mc2 - c2$
Energy = mass
[static equation]

180 Deg (12hr) Noon

90 Deg (6hr) SUNRISE

90 Deg (3 hr)

POSITIVE INFINITY

½ d (r)

0 = 1

~ 7,917.5 miles

(½ m) (c)
magnetized sphere
of mass rotating at
same speed as earth

Equator circumference - 24,900 miles

270 Deg (9 hr)

— Diameter of Earth —

NEGATIVE INFINITY

½ d (r)

(½ m) (c
magnetized sphere
of mass rotating at
same speed as earth

270 Deg (9 hr)

0/360 Deg (24hr) : **Midnight**

The vibrating strings are longer thus greater mass and becomes neutrons, electrons, protons.

[cir. = (pi)((1)(d))(x)]
x = length of vibrating wire

The mass of one cubic centimetre of water at its maximum density at 4 °C = 1 gram/cm3</sup3.

1 gram water = Sphere with diameter
1 Light cm [1.24 cm]

270 Deg (18hr)) SUNSET +3 sec.

Flux field in POSITIVE INFINITY moves in corresponding clockwise

pi = negative infinity minus positive infinity
pi = +1/2 d minus -1/2d
pi = 1 d

Flux field in NEGATIVE INFINITY moves in corresponding counterclockwise

Finalized on Christmas - December 22, 2017,
PRASE GOD

Chapter 4

The Story of Our Life, based on a True Life., to be written by us.

25

CPSIA information can be obtained
at www.ICGtesting.com
Printed in the USA
LVHW071317210319
610688LV00002BA/2/P

9 781545 665367